LESIONS OF CERVIX – A

REVIEW

Dr. Gunvanti Rathod

Dr. Pragnesh Parmar

Dr. Sangita Rathod

Dr. Ashish Parikh

LESIONS OF CERVIX – A REVIEW

Dr. Gunvanti Rathod, MD (Pathology)

Assistant Professor
Department of Pathology
SBKS Medical Institute and Research Centre
Vadodara, Gujarat, India

Dr. Pragnesh Parmar, MD (Forensic Medicine)

Assistant Professor
Department of Forensic Medicine
SBKS Medical Institute and Research Centre
Vadodara, Gujarat, India

Dr. Sangita Rathod, MD (Medicine)

Assistant Professor
Department of Medicine
AMCMET Medical College
Ahmedabad, Gujarat, India

Dr. Ashish Parikh, MD (Medicine)

Consultant Physician
Gayatri Hospital
Gandhinagar, Gujarat, India

DEDICATION

This book is dedicated to my loving daughter **Jayani**.

- **Dr. Gunvanti Rathod**

ACKNOWLEDGEMENTS

We acknowledge the immense help received from the scholars whose articles are cited and included in references of this book. The authors are also grateful to authors / editors / publishers of all those articles, journals and books from where the literature for this book has been reviewed and discussed.

We express our gratitude to our parents and in-laws for their constant encouragement, support and blessings.

It will be an injustice if we do not thank all our students for their innovative ideas and feedback.

CONTENTS

INTRODUCTION

- Gynecological Specimens forms the substantial proportion of workload in histopathological department of most of the laboratories. [1] Complete and accurate assessment of cervical lesion relies on following 3 methods:

    1) Colposcopic examination of cervix
    2) Cervical cytology and
    3) Histopathology of biopsy Specimens.

- Colposcopic examination of cervix with cervical cytology is very frequently used to screen the cervical lesion. Accuracy of cervical cytology is less as compared to Histopathological examination. Therefore histopathological examination of biopsies of cervical lesions and hysterectomy specimens is the gold standard. [2] Carcinoma of the cervix is the most common cancer in Indian women and accounts for 20% of all malignant tumours in the females. [3, 4]

- Worldwide invasive cervical cancer is the second most common female malignancy after breast cancer. [3, 4, 5] Cervical cancer was the most common malignancy in both incidence and mortality among women prior to the 20th century. Today, a dichotomy exists between developing and developed nations. The incidence of cervical cancer in the later has fallen dramatically. [6] The reduction in the incidence of cervical cancer is one of the major public health achievements in developed nations, largely due to the implementation of population-based screening, detection, and treatment programs for preinvasive disease.

- Cervical cancer has different histopathologic types. Squamous cell carcinomas (SCC) account for 75-80% of cervical cancers, adenocarcinoma 15-25%, and adenosquamous carcinomas 3-5%. [7] Adenosquamous tumors exhibit both glandular and squamous differentiation. They may be associated with a poorer outcome than pure SCCs or adenocarcinomas. [8] Occurance of neuroendocrine or small cell carcinomas of cervix is infrequent. Rhabdomyosarcoma is rare and common in adolescents and young women. [9, 10]

- Adenocarcinomas have been rising in incidence since the 1970s; especially in women younger than 35 years of age. [11] Part of the increase may be attributable to an increasing prevalence of HPV infection and part to improvements in screening and prevention of squamous intraepithelial.

- Approximately 80% of cervical cancers occur in developing countries. [3, 4, 5] Worldwide cervical cancer is the fifth most deadly cancer in women. It affects about 16 per 100,000 women per year and kills about 9 per 100,000 per year. [12, 13] Worldwide in 2008, it was estimated that there were 473,000 cases of cervical cancer, and 253,500 deaths per year. [12, 13]

- Cervical cancer is more common in lower socioeconomic groups, in women with early initial sexual activity or multiple sexual partners and in smokers. [14, 15, 16]

- Tumors of cervix are classified by WHO Classification and they are most commonly and widely accepted. [17]

ANATOMY

- The cervix (the term taken from the Latin, meaning neck) is the most inferior portion of the uterus, protruding into the upper vagina. The transition between the endocervix and the lower portion of uterine corpus is termed the isthmus or lower uterine segment [18, 19]

- The cervix measures 2.5 to 3.0 cm in length in the adult nulligravida and when normally positioned is angled slightly downward and backward. The portio vaginalis of the cervix i.e. the part of the cervix protruding into the vagina, also referred to as the exocervix is delimited anteriorly and posteriorly by vaginal fornices. In the centre of the exocervix is the external os which is circular in nulligravida and slit like in the multigravida. The endocervical canal connects the uterus with the vagina. This canal measures 8 mm in diameter and is slightly elliptical. [18, 19, 20]

- The cervix is held in position by two main ligaments that form its principal support. They are –
    1. Uterosacral ligament
    2. The lateral ligament or the transverse cervical ligament or the ligament of Mackenrodlt's. [18, 19, 20]

BLOOD SUPPLY
- The cervix is supplied mainly by the cervicovaginal branch of the uterine artery, in turn a branch of the anterior division of the internal iliac artery. This artery enters the cervix at the level of the isthmus and the terminal branches anastomose with azygous vaginal arteries to form complex network around the cervix. [21, 22, 23] The venous drainage of the cervix is essentially parallel to that of the arterial system. [21, 22, 23]

NERVE SUPPLY

- The cervix is innervated by sympathetic and parasympathetic system derived from pelvic plexus. The endocervix and isthmus contain highest concentration of nerve fibres. Only occasionally do free nerve endings enter the papillae of native squamous epithelium. This may explain the relative lack of painful sensation in the ectocervix. [21, 22, 23]

LYMPHATIC DRAINAGE OF CERVIX

- The lymphatics of the cervix are derived from –
  1. Superficial set arising from cervical mucosa
  2. Deep set of lymphatics arising from the fibromuscular stroma.

- The collecting vessels from the cervix pass laterally within the parametrium to drain into external iliac nodes and posteriorly to the rectal and sacral nodes. [20, 22, 23]

DEVELOPMENT OF THE CERVIX

- The uterus is derived from a pair of mullerian ducts which are formed by invagination of coelomic epithelium during 6th week of gestation. At that time, urogenital sinus and wolffian ducts are already established. [21, 22, 23]

- The two mullerian ducts fuse and then combine with mesonephric and urogenital sinus to form the sinus tubercle. [20, 21] The sinus tubercle gives rise to the vaginal plate epithelium which on its upward migration meets columnar mullerian epithelium at about the 11th week to form the original squamo-columnar junction. [18, 20]

CERVICO VAGINAL EPITHELIUM OF LATE FOETAL LIFE

- In the late foetal life, the original squamo columnar junction is situated just cephalad to the external cervical os. However at term, its position has changed with it now appearing caudal to the external os, usually on the ectocervix or rarely on the vaginalfornix walls. [20]

THE TRANSFORMATION ZONE

- Examination of the cervicovaginal surface in the last trimester of intrauterine life reveals two types of epithelium i.e. columnar and squamous. [18, 19] However there are two types of squamous epithelium. One type is originating with the columnar and lining most of ectocervix and the other is originating in by metaplasic transformation.

- Metaplastic squamous epithelium occurs just cephalad to the original squamo columnar junction. This area is marked by a line called the new or neosquamocolumnar junction. This is found approximately in two thirds of foetal cervices at birth. Between the new squamo-columnar junction and original junction lies the original transformation zone, an area which will persist into adult life and in which occurs dynamic changes throughout adolescent and adult life. [18, 19]

DYNAMIC ANATOMY OF CERVICAL EPITHELIUM

- Three types of epithelium can be identified in the adolescent and adult cervix.

They are –
1. Original or native squamous and columnar epithelium.
2. Metaplastic squamous epithelium.
3. Atypical epithelium comprising pathological epithelia. [18,19]

THE ORIGINAL SQUAMOUS EPITHELIUM
Histological Appearance:

- The original squamous epithelium consists of stratified epithelium. Microscopically it has 5 distinct layers or zone. They are –

- **Zone - 1:** The lowest layer, consisting of single row of small cylindrical cells with relatively large nuclei that are sometimes referred to as basal cells or stratum cylindricum.[18, 19]

- **Zone - 2:** Referred to as parabasal cell layer, prickle cell layer or stratum spinosum profundum. It consists of several layers of polyhedral cells with fairly large nuclei and distinct inter cellular bridges. [18, 19] Mitotic figures can occasionally be found in these two layers.

- **Zone - 3:** Also known as intermediate cell layer, clear cell layer or stratum spinosum. It consists of cells which begin to flatten and have glycogen rich cytoplasm with frequent vacuolation.

- **Zone - 4:** Also called inter epithelial zone or condensation zone is of variable thickness and often not recognizable. It consists of many closely packed polyhedral cells with keratohyaline granular structures. [18, 19]

- **Zone - 5:** Also referred to as the stratum corneum, it consists of superficial layer of cells that are elongated, flattened with small pyknotic nuclei and contain large amounts of cytoplasm. They represent keratinisation and are most plentiful at times of high oestrogenic stimulation. [18, 19]

- The squamous epithelium is separated from the fibrous stoma by a basement membrane that can be demonstrated by an electron microscope.

- The thickness of epithelium depends on hormonal status of the individual. In young girls and elderly women the epithelium is not usually stimulated and is only a few cell layers thick (atrophic epithelium). During the sexually mature period, and as a result of progesterone stimulation, the intermediate cell layer increases and may become very rich in glycogen.

- The superficial layer will also develop under the influence of oestrogen. Occasionally a process of keratinisation is seen above the superficial cells. [18, 19]

THE ORIGINAL COLUMNAR EPITHELIUM
Histological Appearance:

- The endocervix contains glands that connect with the surface. However with the help of three dimensional models, it has been shown that, these so called glands are in reality part of extensive cleft like system.

- These glands are lined by tall, slender and elongated columnar cells, which are uniformly arranged in one layer and closely packed in a cobble stone appearance. The nuclei are round or oval and generally situated in the lower third of the cell. During active secretion as in pregnancy or at ovulation, they are situated at the base of the cells. [18]

- There are two types of columnar cells a.Non ciliated secretory cells b.Kinociliated cells.

- The secretory cells utilize both apocrine and merocrine methods for secretion. Secretory cells stain deeply with PAS during their peak biosynthetic activity. Ciliated cells are covered with kinocilia which beat rhythmically toward the cervical canal and vagina. These cells are more frequent in the endocervix near the endometrial junction. These cells are rarely seen in the ectocervix. These ciliated columnar cells are involved in clearance of secretory macromolecules from the adjacent secretory cells. [18]

METAPLASTIC SQUAMOUS EPITHELIUM

- Squamous metaplasia is seen in 90% of post menarchal cervices and most commonly at the transformation zone. It is a physiological process, occurring late in fetal life, at menarche and during pregnancy. The epithelium once formed does not revert to its original glandular state. The mechanism inducing its development is related essentially to the increasing exposure of columnar epithelium to the acidic vaginal pH. [18, 19, 24] Squamous metaplasia will be dealt in detail in the sections pertaining to metaplastic lesions of the cervix.

THE CERVICAL STROMA

- The connective tissue stroma of cervix is mainly composed of collagen fibres which are dense in the region of the ectocervix and only loosely surround the endocervical glands. Inflammatory infiltrate can be seen deep to the epithelium in normal cervices particularly in the region of transfor mation zone. It has been suggested that inflammatory infiltrate is an immunological response to the cell necrosis and regeneration associated with metaplastic change. [18, 19]

EPIDEMIOLOGY

- Cervical cancer is the leading cancer in India, although breast is leading cancer site globally. It is estimated that in the year 2005, there were about 520,000 cervical cancer cases in the World, of which 443,000 are in the Developing countries. [25] It is the most common cancer in the developing countries.

- In India, it would increase from 0.11 million in the year 2000 to 0.16 million cervix cancer cases in 2010. [25] The proportion ranges from 15% to 55% of females cancers from different parts of the country.

- Over 80% of the cervical cancer present at a fairly advanced stage and around 80,000 deaths are reported due to cervical cancer in India. [25, 26]

AETIOPATHOGENESIS

- One of the earliest epidemiological studies of any cancer was done by Rigoni stern and published in 1842. [27]

- This Italian noted that cervical cancer was found in married women but was virtually absent in celibate groups such as catholic nuns. Other noted that cervical cancer was almost never found in virginal women and that a women's risk of developing cervical cancer was directly related to the number of male sexual partners she had. [27]

- It was also reported that early age of first sexual intercourse, low socioeconomic status, cigarette smoking, early age at first pregnancy, multiple sexual partners increased a women's risk of cervical neoplasia. More obvious that immunosuppression from any cause, including infection with HIV, substantially increased a women's risk of cervical neoplasia. Many epidermiological studies have shown a pattern of cervical cancer that is typical for a STD. For this reason, investigators have focused on etiologic agents that might be passed generally. Virtually every sexually transmitted agent has been studied for its relation to cervical neoplasia, including chlamydia, gonorrhoea, gardenella, mycoplasma, trichomonas and HSV. The herpes simplex virus type-2 received particular attention because it was reported that women whose blood contained antibodies to HSV-2 were at substantially higher risk of having cervical neoplasia compared with patient who were negative for such antibodies. [27]

- Although there is a relation between HSV-2 and virtually all other STDs and cervical neoplasia it seems to be causal relation. In 1976, **Meisels** and **Fortin** [28] and in 1977, **Purola** and **Savia**, [29] reported finding HPV in the nuclei of dysplastic sqamous epithelial cells particularly those that had koilocytic feature. These observations were confirmed by a number of other investigators and it was suggested by **Zur Hausen**, [30] that virus may be important etiologically in cervical carcinogenesis.

- Using electron microscope and antibodies to HPV capsid protein several investigators identified viral particles or HPV related antigen in CIN .they were particularly prevalent in low-grade lesion and become less frequently observed as the cells became less well differentiated . These observations strongly implicated HPV as a possible etiologic agent in cervical neoplasia.

HUMAN PAPILLOMA VIRUS [31, 32, 27]

- HPV are members of a large family of virus known as Papovaviridae .all the viruses in this group are DNA tumor virus. HPV infects virtually all surface epithelia, including skin and mucus membranes. The infected epithelium is characterised by epithelial proliferation at the infected site, by various degrees of epithelial thickening and by papillomatosis.

- Molecular biological studies have demonstrated high level of HPV DNA and capsid antigen, indicating productive viral infection in these koilocytic cells. The HPV genome has been demonstrated in all grades of cervical neoplasia. As the CIN lesions become more severe, the koilocytes disappear, the HPV number decrease , and the capsid antigen disappear, indicating that the virus is not capable of reproducing in less differentiated cells.instead it appears that portion of the HPV DNA become integrated cells.instead it appears that portion of the HPV DNA become integrated into the host cell.

- Integration of the transcriptionally active DNA into the host cells appears to be necessary for malignant growth. Type 16 is the most common HPV type in invasive cancer and in CIN-2 and CIN-3 and is found in 47% of women in both categories, unfortunately, HPV-16 is not very specific, it can be found in 16% of women with low grade lesion and up to 14% of women with normal cytology.

- Other 18, 26, 53 and 66 have been classified as probably high risk and HPV types 6, 11, 40, 42, 43, 44, 54, 61, 70, 72, 81 and CP6108 are included in low risk category. Usually HPV infection does not persist. Most women have no apparent clinical evidence of disease and the infection is eventually suppressed or eliminated.

MOLECULAR PATHOGENESIS OF HPV IN CERVICAL CARCINOMA: [27, 31]

- HPV genome encodes for 6 early (E) proteins, two late (L) proteins which are necessary for gene regulation and cell transformation, regulation of the DNA sequence E6 and E7 act synergistically to transform cells and E7 has the capability to transform cell in isolation. E6 and E7 also interact with P53 and Rb gene causes inhibition of their function as down- regulation in cell replication. E6 and E7 causing disturbance in cell control mechanism and increase in Proliferation of the infected cell rather than entering apoptosis.

- Tumor of cervix is classified by WHO Classification and they are most commonly and widely accepted. [17]

WHO classification of Tumors of cervix [33, 34]

Epithelial tumors	Neuroendocrine tumors
Squamous lesions and precursors	Carcinoid tumor
Squamous cell carcinoma,	Atypical carcinoid tumor
not otherwise specified	Small cell carcinoma
Keratinizing	Large cell neuroendocrine
Nonkeratinizing	carcinoma
Basaloid	Undifferentiated carcinoma
Verrucous	**Mesenchymal tumors and tumor**
Warty	**like conditions**
Papillary	Leiomyosarcoma
Lymphoepithelioma like	Endometrioid stromal sarcoma,
Squamotransitional	low grade
Early invasive (microinvasive)	Undifferentiated endocervical
squamous cell carcinoma	sarcoma
Squamous intraepithelial	Sarcoma botyroides
neoplasia	Alveolar soft parts sarcoma
CIN 3 /SCC in situ	Angiosarcoma
Benign squamous cell lesions	Malignant peripheral nerve sheath
Condyloma acuminatum	tumor
Squamous papilloma	Leiomyoma
Fibroepithelial polyp	Genital rhabdomyoma
Glandular tumors and	Postoperative spindle cell nodule
precursors	**Mixed epithelial and**
Adenocarcinoma	**mesenchymal tumors**
Mucinous adenocarcinoma	Carcinosarcoma
Endocervical	(malignant mullerian mixed
Intestinal	tumor)
Signet ring	Adenosarcoma
Minimal deviation	Wilms tumor
(adenocarcinoma malignum)	Adenofibroma
Villoglandular	Adenomyoma
Endometrioid adenocarcinoma	**Melanocytic tumors**
Clear cell adenocarcinoma	Malignant

Serous adenocarcinoma	melanoma
Mesonephric adenocarcinoma	Blue nevus
Early invasive adenocarcinoma	**Miscellaneous**
Adenocarcinoma in	**tumors**
situ	Tumor of germ cell
Glandular dysplasia	type
Benign glandular lesions	Yolk sac tumor
Mullerian papilloma	Dermoid cyst
Endocervical polyp	Mature cystic teratoma
Other epithelial	**Lymphoid and**
tumors	**hematopoietic tumors**
Adenosquamous carcinoma	Malignant lymphoma (specify
Glassy cell carcinoma variant	type)
Adenoid cystic carcinoma	Leukemia (specify type)
Adenoid basal carcinoma	**Secondary tumors**

BENIGN SQAMOUS CELL LESION

Condyloma accuminata:

- Cervical condyloma is found in less than 6% of women with genital warts on per speculum examination. [35, 36, 37] Characteristically they are multiple. Histologically cervical condyloma shows papillomatosis, acanthosis, parakeratosis and hyperkeratosis. At higher magnification koilocytic change is usually a prominent feature with individual cell keratinisation and multinucleation. There is often inflammatory infiltrate in the underlying cervical stroma.

- Condyloma is frequently associated with CIN changes. [35] Microspectrophotometric technique and immunoperoxidase technique when used to evaluate HPV DNA antigen show that all ordinary cervical condyloma have a diploid nuclear distribution while 59% of condyloma with atypical dysplasia showed HPV antigen. [38]

Squamous papillomas:

- Sqamous papillomas of the ectocervix occur predominantly in young women and are mainly caused by infection with low-risk HPV (LR-HPV) types, such as types 6 and 11. Some may be inverted, hence their surfaces are flat.

- Histologically they consist of thick layers of stratified squamous epithelium with elongated rete pegs that extend deeply into the lamina propria. The basal membrane is intact, the epithelial layers are well differentiated, and acanthosis is usually pronounced. Some lesions may also contain koilocytes in the upper layers. Mitoses are rare.

Fibroepithelial polyp:

- It is more common in the vulvovaginal region, fibroepithelial polyps may occasionally occur in the cervix, usually in young to middle-aged women in their reproductive years. [39] These lesions are typically solitary but may be multiple, particularly during pregnancy.

- Histologically, fibroepithelial-stromal polyps are characteristically polypoid and consist of a prominent connective tissue core with normal to acanthotic overlying epithelium contains a central core of fibrous tissue in which stellate cells with tapering cytoplasmic processes and irregularly shaped thin-walled vessels are prominent features. The most distinctive aspect of these lesions is the stroma, which can range from bland and hypocellular to markedly pleomorphic and hypercellular, the later features producing a particularly worrisome histologic appearance that may mimic a sarcoma. [40]

CERVICAL DYSPLASIA

- Dysplasia was defined for WHO, as a lesion in which, part of the thickness of epithelium is replaced by cells showing varying degree of atypia. The lesions were further graded as mild, moderate and severe dysplasia, although there were no rationally agreed criteria for this grading. [41] Later the term "cervical intraepithelial neoplasia" or "CIN" was introduced which would embrace all grades of dysplasia as well as carcinoma in situ under a single disease heading which more or less accurately convey the morphological and malignant potentials of these lesions.[41]

CERVICAL INTRAEPITHELIAL NEOPLASIA

- Richart in 1967, had suggest the term CIN as a single descriptive term which would embrace all grades of dysplasia as well as carcinoma in situ,under a single disease heading and accurately convey the morphological unity and malignant potential of these lesion. However this terminology is best utility for histological diagnosis. In 1982 Buckley et al and Fecenxzy recongnized three grades of CIN.

1- CIN –I corresponding to mild dysplasia
2- CIN-II corresponding to moderate dysplasia
3- CIN-III corresponding to severe dysplasia

The histological features taken into account while accessing CIN are:

1. Differentiation (maturation, stratification)
    a. Presence or absence
    b. Proportion of epithelium showing differentiation.
2. Nuclear abnormalities
    a. N:C ratio
    b. Hyperchromasia
    c. Nuclear pleomorphism and anisokaryosis.

3. Mitotic activity
    a. Number of mitotic figures
    b. Height in epithelium
    c. Abnormal configuration.

- Taking these general features into account, the histopathology of various grades of CIN is as follows. [41]

CIN - I: Maturation is normal in the upper two thirds of epithelium, although slight nuclear atypia persist upto the surface, representing a delay in nuclear maturation. Nuclear abnormalities are most marked in the basal third of epithelium. Mitotic figures are present but not numerous and are also confined to the basal third of epithelium. Abormal mitotic forms are very rare. [42]

CIN - II: Maturation is normal in the upper half of epithelium. Nuclear abnormalities are more marked than CIN I and extend further through the epithelium. Occasional abnormal mitotic figures may be seen, but still confined to the basal third of epithelium.

CIN - III: Maturation is absent and nuclear abnormalities are marked throughout the thickness of epithelium. Abnormal mitotic figures are frequent and seen at all levels of epithelium. CIN III resembles carcinoma in situ. [42]

Another system reduces these entities to two: LSILs and HSILs. [43]

Low-grade squamous intraepithelial lesions:
- Histologically, these lesions have cytologic atypia in a distribution characteristic of flat or exophytic condylomata or their immature variants Exophytic LSILs (condylomata acuminata) exhibit thickening of the epithelium (acanthosis) with variable papillomatosis and viral cytopathic effect (koilocytotic atypia) in the middle and upper portions of the epithelium. Koilocytotic atypia is a constellation of cellular changes that include variation in nuclear size and shape, wrinkled nuclei, hyperchromasia, binucleation or multinucleation, and perinuclear halos. [44] Nuclear atypia is minimal in the lower half of the epithelium, there is a low mitotic index, and bizarre or abnormal mitoses are absent.

- Most exophytic condylomata are associated with HPV-6 or HPV-11 Flat LSILs (flat condylomata) have a similar appearance but lack the papillary architecture of an exophytic lesion Approximately 60% to 70% of flat LSILs are associated with high-risk HPV types; 10% to 15% are associated with HPV-16. [45]

- Immature condyloma or papilloma that is characterized by a proliferation of immature squamous cells with mild cytologic atypia, frequently associated with more mature areas of typical condyloma, and a tendency to extend into the endocervical canal.[46, 47, 48] Immature condylomata are characteristically positive for HPV-6 or HPV-11.

High-grade squamous intraepithelial lesions:
- HSILs share several morphologic features, specifically nuclear atypia in all layers of the epithelium, at least in a portion of the lesion. [49] Lesions with koilocytotic atypia or maturation

conform to the image of CIN II, whereas those without discernible maturation fall into the category of CIN III.Three histologic variants of HSIL-

1- Keratinizing or koilocytotic HSIL is characterized by prominent superficial keratinization in addition to atypia. Nuclei in a keratinizing HSIL exhibit much greater variation in contour, dense or coarse-appearing chromatin, and often, intensely eosinophilic cytoplasm.

2- HSILs with immature metaplastic differentiation-These include a rounded or undulating epithelial-stromal interface, uniform-appearing immature cells with variable Cytoplasmic differentiation.

3- Stratified mucin-producing intraepithelial lesion- These lesions are morphologically similar to immature metaplastic HSILs, with the exception of conspicuous mucin droplets.

- In one study, stratified mucin-producing intraepithelial lesions were associated with a greater risk of coexisting invasive carcinoma. [50] A follow-up of various grades of CIN suggest that more than 40% cases of CIN III progress to invasive carcinomas and hence the triad of pap smear, colposcopy and biopsy are mandatory in the evaluation of unhealthy cervix.[51]

- HPVs, particularly high risk HPVs, are associated with alterations in the cellcycle. Expression of a generic cell cycle proliferation marker (Ki-67) is typically confined to the suprabasal cells of the lower third of the normal epithelium. The presence of Ki-67 positive cells in the upper epithelial layers occurs in HPV infection, which induces cell cycle activity in these cells. [52, 53] P16ink4a, a cyclin dependent kinase inhibitor, is a promising marker of CIN. [54, 55]

MICRO INVASIVE CARCINOMA: [57, 58]

- Approximately 4 to 7% of CIN lesion has been associated with superficial invasion. [58, 59] They represent the preclinical stages of invasive cervical carcinoma. A microinvasive carcinoma (MIC; FIGO stage IA, TNM 1a1 and 1a2) was originally defined as an early invasive carcinoma that could not be detected grossly, only histologically.

- Microinvasion often develops from precancerous epithelium located on the surface, as well as from that located in the glands According to the 1985 modification of FIGO staging, a MIC was defined as a carcinomatous invasion not exceeding 5 mm in depth and 7mm in horizontal spread. Sites of microinvasion are often surrounded by dense lympho-plasmocytic infiltrates. In recent years, the proportion of invasive cervical carcinoma that invaded less than 5 mm in depth at diagnosis has increased more than 10-fold and currently is approximately 21%. [61, 62]

SQUAMOUS CELL CARCINOMA AND ITS VARIANTS:

- It is commonest malignant tumor Accounts for 90% of all malignant neoplasm of cervix. Most squamous cell carcinomas evolve from a precancerous lesion and persistent infection by HPV. [62, 63] From 10% to 20% of CIN 3 lesions progress to cancer if they are untreated, and the time course for this evolution has been estimated to range from 3 to more than 20 years.[63, 64]

- Squamous cell carcinomas have been classified according to the degree of squamous differentiation (grades I–II) or according to cell type. Reagan and colleagues [65] subdivided squamous cancer of the cervix into (a) Large cell keratinizing carcinoma (b) Large cell nonkeratinizing carcinoma, and (c) Small cell nonkeratinizing carcinoma

A- Large Cell Keratinizing Squamous Cell Carcinoma:

- It represents about 25%. [66] Presence of whorls of neoplastic sqamous cells with large nuclei and keratinized cytoplasm. There is Formation of intercellular bridge. Dyskeratosis is present. Cells are having abundant eosinophilic cytoplasm some cells, with clear cytoplasm, containing glycogen are present giving PAS positive reaction. Mitotic figures and cellular pleomorphism are present. Keratohyaline granule is present.

B- Large Cell Nonkeratizing Squamous Cell Carcinoma:

- It represents about 70%. [66]   The tumor shows large polygonal cells with eosinoplihic cytoplasm, keratiziation is absent or sparse. Presence of anisonucleosis, nuclear pleomorphism, high N: C ratio, hyperchromatic nuclei, prominent nucleoli frequently observed. Mitosis is more in number. Cell occurs in loose uniform cluster. Intercellular bridges are poorly defined.

C-Small Cell Non Keratinizing Carcinoma:

- It represents approximately 5%. [66] Tumor cells are arranged in large aggregates or sheets. The tumor shows small cell with sparse cytoplasm, with high N: C ratio. Hyperchromatic nuclei and frequent mitotic figure represent with absent or sparse keratinization.

OTHR MICROSCOPIC VARIANTS
Verrucous Carcinoma:

- Pure verrucous carcinoma of the cervix is a very rare lesion. [67] Composed of very well-differentiated keratinizing squamous cells growing in frondlike papillae. Stromal invasion is always bulky, lymphatic involvement or distant spread is rare. This tumor is more common in the vulva and vagina. These carcinoma are CEA-negative Verrucous carcinoma has a tendency to recur locally, and the therapy of choice is wide local excision.

Lymphoepithelioma-Like Carcinoma:

- These extremely uncommon variants are characterized by tumor cells arranged in discohesive clusters in a background of marked inflammation .Epstein–Barr virus has been detected in some of these carcinomas.

Warty (Condylomatous) Carcinoma:

- The warty (condylomatous) type is a squamous cell carcinoma with obvious cellular signs of HPV infection, e.g., koilocytosis, and with a warty surface High risk HPV-DNA is typically detected. [68] It is also referred to as condylomatous squamous cell carcinoma.

Basaloid Squamous Cell Carcinoma:

- Basaloid squamous cell carcinoma is composed of nests of immature, basal type squamous cells with scanty cytoplasm that resemble closely the cells of squamous carcinoma in situ of the cervix.

Papillary Squamous Cell Carcinoma:

- In 2000 Brinck et al describe Papillary squamous cell carcinoma is another rare variant of cervical squamous cell carcinoma, presenting the gross and histological appearance of a benign papilloma undergoing dysplastic cellular change. Beneath the papillary dysplasia, strands of invasive carcinoma can be detected.

Squamo-Transitional Cell Carcinoma:

- Rare transitional cell carcinomas of the cervix have been described that are apparently indistinguishable from their counterpart in the urinary bladder. The detection of HPV type 16 and the presence of allelic losses at chromosome 3p with the infrequent involvement of chromosome 9 suggest that these tumours are more closely related to cervical squamous cell carcinomas than to primary urothelial tumours. [69, 70]

Spindle Cell (Sarcomatoid) Carcinomas:

- Uncommon variant that is typified by a squamous carcinoma that undergoes spindle cell differentiation these tumors, like lymphoepithelial-like tumors, are HPV positive. [71]

BENIGN GLANDULAR LESIONS

Cervical Polyps:

- Endocervical Polyps are relatively innocuous inflammatory non-neoplastic tumours that occur in 2-5% of adult women. [72] They are not true neoplasms but probably the result of chronic inflammatory changes (chronic polypoid cervicitis) and most often arises within endocervical canal. Their size varies from small and sessile to large mases that may protrude through cervical os. [73] Aridogan et al [74] suggested multiparity, chronic cervicitis, foreign bodies, and unpredictable oestrogen secretion as aetiological factors causing the development of cervical polyps.

- Cervical polyps are composed of mature endocervical epithelium and are considered as focal, hyperplastic protrusions of endocervical folds including the epithelium and substantia propria, rather than true neoplasms. Cervical polyps are usually found during the 4th to 6th decades and commonly in multigravidas. Cervical polyps are usually have no symptom but rarely may cause profuse leucorrhoea and abnormal bleeding.

- Grossly most polyps occur singly and measure a few millimeters to 2-3 cm. In rare instances they may reach gigantic proportions protruding beyond the introitus resembling a carcinoma. [75, 76, 77, 78]

- Microscopically cervical polyps display a variety of patterns that vary according to the preponderance of one or another of tissue components. They are [75]
  1. Endocervical mucosal polyp – most common consists of mucinous epithelium that lines crypts, with or without cystic changes.
  2. Fibrous polyp – consists of overgrowth of connective tissue stroma of cervical portion.

3. Vascular polyp – blood vessels predominate and in these polyps, squamous metaplasia is a frequent feature, along with dense chronic inflammatory infiltrate.
4. Mixed polyps – polyps originating in the isthmus often have an admixture of endocervical and endometrial type and referred to as mixed polyps.
5. Inflammatory polyp – consists of dense chronic inflammatory infiltrate, and occasionally it is so dense, that only granulatio n tissue predominates without a lining epithelium.
6. Decidual polyps – seen in pregnancy, rarely decidualisation of endocervical stroma produces a polypoid protrusion from the endocervix.
7. Mesodermal stromal polyp – (pseudosarcoma botryoides) are benign exophytic proliferations of stroma and epithelium that can occur in the vagina and cervix of women of reproductive age group and more commonly in pregnancy. The polyp consists of edematous stroma usually composed of bland appearing plump fibroblasts. Occasionally, multinucleate stromal giant cells bizarre fibroblasts with irregular hyperchromatic nucleus can be seen. They are mistaken for sarcoma botryoides. A careful search will exclude mitosis, rhabdomyoblasts and cambium layer, which will differentiate it from malignancy.

Müllerian Papilloma:

- This rare lesion, seen almost exclusively in children, presents as a papillary excrescence on the ectocervix or vagina. [79] Microscopically, it contains small papillae lined by a single layer of cuboidal epithelium without nuclear atypia or significant mitotic activity.

GLANDULAR DYSPLASIA

- A glandular lesion characterized by significant nuclear abnormalities that are more striking than those in glandular atypia but fall short of the criteria for adenocarcinoma in situ. Nuclear hyperchromasia and enlargement identify the involved glands, and pseudostratification of cells is prominent. Cribriform and papillary pattern formations are usually absent.

ADENOCARCINOMA IN SITU

- ACIS was first described in 1953. Average age at presentation is 35 to 40 years, approximately 10 years earlier than invasive adenocarcinoma. [80, 81, 82, 83, 84, 85] 30% to 60% of cases are associated with an SIL. More than 90% of cases of adenocarcinoma and ACIS have detectable HPV. [86] ACIS arises either from columnar epithelium or, more likely, reserve cells with the capacity to undergo columnar cell differentiation The classic histologic picture of ACIS includes both architectural and epithelial components The most common presentation consists of closely arranged clusters of glands with a distinctly higher intensity of chromasia than the surrounding crypt epithelium Histologically-the lining epithelium is stratified and crowded, and it consists of moderately enlarged nuclei with coarse chromatin . Mitotic figures are easily found, Nucleoli are usually small and inconspicuouspresence of apoptotic bodies, found in 80% of cases. [87] The most common subtype of ACIS is the endocervical type, other pattern-endometrioid pattern, intestinal type.

EARLY INVASIVE ADENOCARCINOMA

- In early invasive adenocarcinoma, microinvasion must be limited to 5 mm in depth with extension beyond the normal glandular field. Early invasive adenocarcinoma refers to a glandular neoplasm in which the extent of stromal invasion is so minimal that the risk of local lymph node metastasis is negligible Lesions should also be less than 7.0 mm in length and should be devoid of capillary-lymphatic space invasion. [88, 89, 90, 91]

- Histopathology: The sine qua non of microinvasive adenocarcinoma is stromal invasion. There may be marked glandular irregularity with effacement of the normal glandular architecture, the tumour extending beyond the deepest normal crypt. There may be a stromal response in the form of oedema, chronic inflammatory infiltrate or a desmoplastic reaction. Lymphatic capillary-like space involvement is helpful in confirming invasion.

ADENOCARCINOMA OF THE CERVIX

- All types of adenocarcinomas of the endocervix have become more common over the past three decades especially in young women.

- Adenocarcinomas account for approximately 15% of invasive cervical cancers, but in some series the percentage is as high as 25%. [93] Great majority of endocervical adenocarcinomas contain HPV 18 or 45. The prognosis of all types of endocervical adenocarcinomas is generally less favourable than that of squamous carcinomas.

- Risk factors for adenocarcinoma are increased numbers of sexual partners, high rates of HPV infection, [86] prolonged use of birth control pills. [94, 95, 96] Gross examination, advanced cervical adenocarcinoma may present as an exophytic mass, an ulcerated plaque, or diffuse cervical enlargement (barrel-shaped cervix). [97] Adenocarcinomas of the cervix more frequently metastasize to the ovaries, upper abdomen, or distant organs than do squamous carcinomas. [98]

MUCINOUS ADENOCARCINOMA

- It is an adenocarcinoma in which at least some of the cells contain a moderate to large amount of intracytoplasmic mucin. The mucinous (endocervical) type of adenocarcinoma is most frequently encountered. All degrees of differentiation can be found. Most of the mucinous adenocarcinomas are well or moderately differentiated. Their interbranching glands may show microglandular changes and contain variable amounts of intracytoplasmic mucin. All types of mucinous carcinomas, regardless of their degree of differentiation, almost always stain positively for CEA.

Endocervical variant:
- Approximately 70% of cervical adenocarcinomas are of the endocervical (mucinous) type. [97] The tumor may contain simple, complex, or dilated glands, solid sheets, cords, or single cells.

Intestinal variant:
- These tumours resemble adenocarcinoma of the large intestine. Intestinal-type change may be found diffusely or only focally within a mucinous tumour. They frequently contain goblet cells and less commonly endocrine and Paneth cells.

Signet-ring cell variant:
- Primary signet-ring cell adenocarcinoma is rare in pure form. [99] Signet-ring cells occur more commonly as a focal finding in poorly differentiated mucinous adenocarcinom and adenosquamous carcinomas.

Adenoma Malignum (Minimal Deviation Adenocarcinoma):
- This well-differentiated subtype accounts for approximately 1% to 2% of all endocervical adenocarcinomas. [100] It has been associated with Peutz-Jeghers syndrome. [101] Adenoma malignum usually results in a diffusely enlarged, barrel-shaped cervix and may cause a vaginal discharge adenoma malignum is suspected on the basis of the infiltrative appearance of cystically dilated, irregular, or "claw-shaped" glands that permeate the endocervical stroma. The cells lining the glands are extremely well differentiated, with basally located nuclei and abundant, tall columnar cytoplasm. The nuclei are slightly larger than those of normal endocervical cells and may have a small eosinophilic nucleolus. A range of nuclear atypia, with slightly more enlargement and hyperchromasia in some foci and the presence of occasional mitotic figures, is helpful for diagnosis.

Villoglandular Adenocarcinoma:
- It is the variant of endocervical-type adenocarcinoma, with a prominent papillary growth pattern characterized by narrow or

broad papillary cores with a spindled or inflamed stroma. Adjacent ACIS is often present, lymph node metastases are rare.

ENDOMETRIOID ADENOCARCINOMA

- The second most frequent category of pure cervical adenocarcinoma is the endometrioid type.The endometrioid type of adenocarcinoma may arise from metaplasia or from ectopic endometrial glands, representing embryological remnants displaced in the deeper portions of the cervical wall. Its histological structure is like that of the endometrial-type adenocarcinoma arising from the corpus mucosa. PAS-positive and diastase-resistant secretions may be found in the glandular lumina, but not in the cytoplasm. Although most reported cases have had a benign course, one death was reported. [102]

CLEAR CELL CARCINOMA

- It develops in young women who have been exposed in utero to diethylstilbestrol. [103, 104] and it develops in older age group without exposure. [107] the clear cell adenocarcinoma may be mainly solid or glandular with papillary protrusions, hobnail cells, and a glycogen-rich cytoplasm. The endocervical clear cell adenocarcinomas most likely arise from reserve cells, which presumably through faulty differentiation remain at an intermediary stage of development between incomplete keratinization and the secretion of mucins.

SEROUS ADENOCARCINOMA

- Dallenbach-Hellweg and Poulsen in 1996 describe the serous adenocarcinoma, too, closely resembles those originating in the ovaries and endometrium. A complex pattern of papillae with cellular budding and the frequent presence of psammoma bodies characterize serous adenocarcinoma. It grows as a papillary proliferation with broad or thin fibrovascular cores, lined by pleomorphic epithelial cells with large nuclei, prominent nucleoli, and many mitotic figures.

MESONEPHRIC ADENOCARCINOMA

- The mesonephric adenocarcinoma is a rare cervical tumor, representing approximately 3% of the adenocarcinomas of this location. The true mesonephric adenocarcinoma is a well-defined tumor,which originates from remnants of the mesonephric duct in the lateral wall of the cervix .The histologic patterns are quite variable and include the ductal, retiform, tubular, solid, and sex-cord–like types. [106] The nuclei of mesonephric carcinomas are usually of low to moderate grade. Mucin is absent in the sparse cytoplasm, but a small amount of PAS-positive material may be detected in the glandular lumina Most cases are reactive for keratin, epithelial membrane antigen, vimentin, and CD10. [107, 108]

OTHER EPITHELIAL TUMORS

ADENOSQUAMOUS CARCINOMA

- The incidence of adenosquamous carcinoma has significantly increased during the past decades, particularly in younger women. Since it originates from the bipotential reserve cells, and it is frequently associated with HPV infection type 18 or the equally frequent type 16. Adenosquamous carcinomas have been said to account for between 15% and 35% of all cervical carcinomas with a glandular component, Many squamous cell carcinomas without an obvious glandular component stain positively for mucin. [109] Adenosquamous carcinomas have been reported to have a worse prognosis than other adenocarcinomas. [110, 111]

GLASSY CELL CARCINOMA

- Glassy cell carcinoma is considered a variant of adenosquamous carcinoma in which the differentiating features are recognizable only by electron microscopy. [112] this tumor is characterized by a solid growth of large tumor cells with abundant eosinophilic (ground-glass) cytoplasm. Nuclei are large with prominent eosinophilic nucleoli, and the stroma contains an inflammatory infiltrate, often with many eosinophils.

ADENOID BASAL CARCINOMA

- It occurs in elderly women, does not grow as a visible mass, and is associated with an HSIL of the surface epithelium. [113] It consists of small, round to oval branching nests of cells, resembling those of basal cell carcinoma of the skin, with palisading of the peripheral cell layers. There are no or only rare cystic pseudoglands. The tumor cells are small and uniform, with rounded hyperchromatic nuclei in scanty cytoplasm. Mitoses are infrequent. These tumors may demonstrate four distinct components. The first is an HSIL, second is an invasive squamoid component with prominent central maturation and atypia and a discrete rim of basal cells, the third component, composed of small basaloid islands, extends peripherally. A fourth component, composed of small acini (adenoid) that reflect clear-cut columnar cell differentiation

ADENOID CYSTIC CARCINOMA

- This rare tumor occurs in older women, it presents with a cervical mass. It grows predominantly as nests of cells in a cribriform pattern with intervening hyaline cores, resembling adenoid cystic carcinoma of the salivary glands. [113] It is composed of fairly uniform, small basaloid cells with scanty cytoplasm and rounded or angulated hyperchromatic nuclei. These cells forms small branching or larger strands, surround small or larger cysts , or form small acini filled with a hyaline or basement membrane-like material rich in acid mucopolysaccharides. This tumor is often associated with foci of squamous cell or adenocarcinoma. Adenoid cystic carcinomas show early lymphatic invasion and are more aggressive than most cervical adenocarcinomas, along with poor prognosis.

NEUROENDOCRINE TYPE

- This group includes carcinoids, atypical carcinoids, and small cell and large cell neuroendocrine carcinomas. [114] Neuroendocrine carcinomas most likely develop from neuroendocrine cells occurring in the normal endocervix or from stimulated multipotential reserve cells of the endocervical epithelium undergoing neuroendocrine metaplasia and hyperplasia . Almost all neuroendocrine carcinomas of the cervix are associated with HPV 18 or seldom HPV 16.

Carcinoid tumors:

- Carcinoid tumors originate from endocervical argyrophil cells. The tumors consist of rather uniform, small, round or ovoid cells growing in solid sheets, small lobules, or trabeculae. They are usually benign.

Atypical carcinoids:

- It shows cytologic atypia and contains foci of necrosis. Mitoses are frequent. Mitotic activity (5-10 per high power field) vascular invasion is often obvious, giving evidence of the malignant behavior of these tumors. [114]

Small cell carcinomas of neuroendocrine origin:

- Small cell carcinomas account for 1-6% of cervical carcinomas. [115] the 5 year survival rate is reported to be 14-39% [115] are considered to be the poorly differentiated variety of carcinoid tumors, resembling small cell carcino mas of the lung, although mutations of p53 and LOH are less common in the cervical tumor- The cells of these tumors are spindle shaped. Mitoses are numerous. Their scanty cytoplasm contains argyrophilic neurosecretory granules, demonstrable by the Grimelius stain, S100, or by specific neuroendocrine markers such as NSE, chromogranin A, and synaptophysin. Among these, CD 56, a neural cell adhesion molecule, appears to be the most sensitive marker.

Large cell neuroendocrine carcinomas:

- These are very rare. Their cells contain large nuclei with prominent nucleoli in abundant cytoplasm. Mitoses are very frequent. The cells are positive for cytokeratin, chromogranin and focally contain mucin. Most of these tumors contain foci of in situ or invasive adenocarcinoma.

UNDIFFERENTIATED CARCINOMA

- Undifferentiated carcinoma is a carcinoma lacking specific differentiation. The differential diagnosis includes poorly differentiated squamous cell carcinoma, adenocarcinoma, glassy cell carcinoma and large cell neuroendocrine carcinoma.

MESENCHYMAL TUMORS

- A variety of rare benign and malignant mesenchymal tumours that arise in the uterine cervix and which exhibit smooth muscle, skeletal muscle, vascular, peripheral nerve and other types of mesenchymal tissue differentiation. Smooth muscle tumours are the most common.

LEIOMYOSARCOMA

- A malignant tumour composed of smooth muscle cells. It presents as a mass replacing and expanding the cervix or as a polypoid growth. The tumours have a soft and fleshy consistency and often contain areas of necrosis or haemorrhage.Leiomyosarcomas show hypercellular interlacing fascicles of large spindleshaped or round cells with diffuse moderate to marked nuclear atypia, a high mitotic rate, atypical mitoses, single or multiple prominent nucleoli and tumour cell necrosis. Infiltrative borders and vascular invasion are also frequently seen.

ENDOMETRIOID STROMAL SARCOMA

- A sarcoma arising outside of the fundus composed of cells resembling endometrial stromal cells. This tumour may arise from cervical endometriosis.

UNDIFFERENTIATED ENDOCERVICAL SARCOMA

- These are characterized by a polypoid or infiltrative cervical growth similar to that exhibited by malignant peripheral nerve sheath tumours arising in the uterine cervix. [116, 117] The tumour is composed of spindle or stellate-shaped cells with scanty cytoplasm, ill defined cell borders and oval hyperchromatic nuclei arranged in a sheet-like, fasciculated or storiform pattern. [116] Nuclear atypia and markedly increased mitotic activity are seen in all cases, as well as areas of haemorrhage, necrosis and myxoid degeneration.

SARCOMA BOTRYOIDES

- Usually grows in a polypoid fashion. The polypoid masses have a glistening translucent surface and a soft consistency and may be pedunculated or sessile. Their size ranges from 2-10 cm. [118] The sectioned surface of the tumour appears smooth and myxoid with small haemorrhagic areas. A tumour composed of cells with small, round, oval or spindle-shaped nuclei, some of which show evidence of differentiation towards skeletal muscle cells. The use of neoadjuvant chemotherapy allows a more conservative approach for these neoplasms. [119]

ALVEOLAR SOFT PART SARCOMA

- A sarcoma characterized by solid and alveolar groups of large epithelial-like cells with granular, eosinophilic cytoplasm. These appear macroscopically as a polyp or an intramural nodule measuring less than 5 cm and have a friable or solid consistency. Most of the tumors exhibit an alveolar architecture, where nests of tumour cells with central loss of cellular cohesion are supported by thin-walled, sinusoidal vascular spaces. The tumor cells have an abundant eosino-philic cytoplasm, large nuclei, prominent nucleoli and contain PAS-positive, diastase-resistant, rodshaped crystals. [120] Alveolar soft part sarcomas of the female genital tract, including those primaries in the uterine cervix, appear to have a better prognosis than their counterpart in other sites.[121]

ANGIOSARCOMA

- It is a malignant tumour the cells of which variably recapitulate the morphologic features of endothelium. The macroscopic appearance of angiosarcoma is similar to that in other sites form ingahaemorrhagic, partially cysticornecrotic mass, [122] and the neoplastic cells are immuno reactive for CD31, CD34, and factor VIII-related antigen. [122]

MALIGNANT PERIPHERAL NERVE SHEATH TUMOR

- It is a malignant tumour showing nerve sheath differentiation. The tumour is composed of fascicles of atypical spindle cells invading the cervical stroma and surrounding endocervical glands with a pattern reminiscent of adenosarcoma. Myxoid paucicellular areas are characteristically intermixed with others with a dense cellularity. [123] Mitoses are common. The tumour cells are positive for S-100 protein and vimentin and negative for HMB-45, smooth muscle actin, desmin and myogenin. [124]

LEIOMYOMA
- A benign tumour composed of smooth muscle cells. Leiomyoma is the most common benign mesenchymal tumour of the cervix. It has been estimated that less than 2% of all uteri contain cervical leiomyomas, and that about 8% of uterine leiomyomas are primary in the cervix. [125, 126]

GENITAL RHABDOMYOMA
- A rare benign tumour of the lower female genital tract composed of mature striated muscle cells separated by varying amounts of fibrous stroma the tumor is composed of rhabdomyoblasts with small, uniform nuclei dispersed in a myxoid and oedematous stroma. The typical cambium layer of sarcoma botryoides is absent. [127]

POSTOPERATIVE SPINDLE CELL NODULE
- A localized, non-neoplastic reactive lesion composed of closely packed proliferating spindle cells and capillaries simulating a leiomyosarcoma occurring at the site of a recent excision. The lesion develops at the site of a recent operation several weeks to several months postoperatively. [128, 129] The lesion is composed of closely packed, mitotically active, spindle-shaped mesenchymal cells and capillaries often with an accompaniment of inflammatory cells, and may infiltrate the underlying tissue.

MIXED EPITHELIAL AND MESENCHYMAL TUMORS
- Tumors composed of an admixture of neoplastic epithelial and mesenchymal elements. Each of these components may be either benign or malignant. Most commonly involve elderly

postmenopausal women. [130] The presenting symptom is usually abnormal uterine bleeding.

CARCINOSARCOMA

- Carcinosarcomas most commonly involve elderly postmenopausal women. [130] HPV infection, especially HPV 16, has been found in the epithelial and mesenchymal components suggesting a role in the evolution of these neoplasms. [131] The histological features are similar to its counterpart in the uterine corpus. However, the epithelial elements are more commonly non-glandular in type and include squamous (keratinizing, non-keratinizing or basaloid), adenoid cystic, adenoid basal or undifferentiated carcinoma. [130, 131] Adjacent severe dysplasia of the squamous epithelium has also been described. Cervical carcinosarcomas are aggressive neoplasms, and treatment is usually radical hysterectomy followed by chemotherapy and or radiotherapy.

ADENOSARCOMA

- Cervical adenosarcomas are much less common than their counterparts in the uterine corpus. A neoplasm composed of an admixture of benign epithelial and malignant mesenchymal elements. Adenosarcomas may or may not invade the underlying cervical stroma. The therapy is usually simple hysterectomy, and radiation may be considered for deeply invasive neoplasms.

WILMS TUMOR

- It is a malignant tumor showing blastema and primitive glomerular and tubular differentiation resembling Wilms tumour of the kidney. Within the cervix have been described, usually present in adolescents. [132, 133]

- Macroscopically, these neoplasms are composed of polypoid masses that protrude through the vagina. Histologically, the classic triphasic patte rn of epithelial, mesenchymal and blastemal elements may be present. In two cases prolonged survival has been reported following local excision and chemotherapy. [132, 133]

ADENOFIBROMA

- These are uncommon in the cervix and are more commonly found within the uterine body. [134] Cervical adenofibromas are polypoid neoplasms that usually protrude into the endocervical canal. On sectioning small cystic spaces may be identified.

- Histologically, adenofibroma is a benign papillary neoplasm composed of fronds lined by benign epithelium that is usually glandular in type. The epithelium may be cuboidal, columnar, attenuated, ciliated or mucinous. The mesenchymal component shows little mitotic activity and is usually composed of non-specific fibrous tissue the therapy is usually local excision or simple hysterectomy.

ADENOMYOMA

- A tumor composed of a benign glandular component and a benign mesenchymal component composed exclusively or predominantly of smooth muscle. These tumors are rare within the cervix.

- Cervical adenomyomas are usually polypoid lesions with a firm sectioned surface. Three variants of cervical adenomyoma have been described the endocervical type, the endometrial type and atypical polypoid adenomyoma.

- Simple polypectomy or local excision cures most cervical adenomyomas.

MELANOCYTIC TUMORS

MALIGNANT MELANOMA

- A malignant tumour of melanocytic origin. All occurred in adults, and approximately one-half had spread beyond the cervix at the time of presentation. [135] and these tumors commonly present with abnormal vaginal bleeding.

- Malignant melanomas are typically described as polypoid or fungating, pigmented masses. The histological appearance of cervical melanomas is noteworthy for the frequent presence of spindle-shaped cells. Desmoplastic and clear cell variants have also been reported. [136] The prognosis for patients with cervical melanoma is dismal, with only two reports of patients surviving more than 5 years. [137]

BLUE NAEVUS

- A naevus composed of dendritic melanocytes that are typically heavily pigmented. Benign pigmented lesions are asymptomatic and are typically incidental findings in hysterectomy specimens. [138, 139] As most blue naevi occur in the endocervical canal, they are not visible colposcopically. [138, 139]

- Blue naevi are recognized histologically by the presence of poorly circumscribed collections of heavily pigmented, bland, spindle-shaped cells with fine dendritic processes in the superficial cervical stroma. They are most commonly located under the endocervical epithelium, but examples that involved the ectocervix have been reported. [140]

MISCELLANEOUS TUMORS

TUMORS OF GERM CELL TYPE
Yolk sac tumor:
- A primitive malignant germ cell tumour characterized by a variety of distinctive histological patterns, some of which recapitulate phases in the development of the normal yolk sac. The cervix is the second most common site in the lower female genital tract for yolk sac tumour after the vagina. These tumours commonly present with abnormal vaginal bleeding. Yolk sac tumours are polypoid, friable masses, protruding into the vagina. [141] the prognosis for patients with cervicovaginal yolk sac tumours is good with modern chemotherapy

Dermoid cyst:
- A mature teratoma characterized by a predominance of one or a few cysts lined by epidermis accompaned by its appendages. Cervical teratomas appear as smooth cervical polyps that may be pedunculated. [143] The histological appearance is indistinguishable from mature teratomas at other sites. Glial and squamous epithelial elements are common, but a wide range of mature tissue types have been reported. [143]

LYMPHOID AND HEMATOPOIETIC LYMPHOMA AND LEUKAEMIA
- It is a malignant lymphoproliferative or haematopoetic neoplasm that may be primary or secondary. Involvement of the cervix by lymphoma or leukaemia may rarely be primary but is more commonly part of systemic disease with no specific symptoms referable to the cervix. [144] Cervical involvement by lymphoma or leukaemia the cervix appears enlarged and barrel-shaped, although polypoid or nodular masses may be seen. [144]

SECONDARY TUMORS [33, 34]

- Pelvic carcinomas, mainly those arising in the endometrium, ovary, rectum, and bladder, may extend into the cervix. Distinguishing metastatic tumors from primary endocervical adenocarcinoma is often possible immunohistochemically. Metastases from distant primary tumors are rare. If they occur, the primary site most likely will be the breast or the gastrointestinal tract.

- Carcinomas of the female genital tract particularly cancer of the cervix, account for almost 12% of all cancers in women, and so represent the second most frequent gynaecological malignancy in the world. [145] Cancer of cervix accounts for 470,000 new cases of all cancer each year in the world. [146]

- Cervical cancer is the third largest cause of cancer mortality in India after cancers of the mouth & oropharynx, and oesophagus, accounting for nearly 10% of all cancer related deaths in the country. India has a disproportionately high burden of cervical cancer. Cervical cancer was the third largest cause of cancer mortality in India.

- In India 90,000 of new cases of cervical cancer occurs every year. [147] Cancer that develops in the ectocervix is usually squamous cell carcinoma, around 80- 90% of cervical cancer cases (more than 90% in India) are of this type. [147]

- Cancer that develops in the endocervix is usually adenocarcinoma. In addition, a small percentage of cervical cancer cases are mixed versions of the above two and are called adenosquamous carcinomas or mixed carcinomas.

Table - 1: Proportion of occurance of SCC and Adenocarcinoma

Study (year)	Squamous cell carcinoma	Adenocarcinoma
Haghdel et al (1999) [148]	88%	11%
Smith et al (2000) [149]	88%	12%
M.A.Ijaiya et al (2004) [150]	85.2%	5.4%

Table - 2: Mean age of patient and cervical carcinoma

Histology type	Balkachew nigatu et al study (Mean age years) [151]	Dhakal HP et al study (Mean age years) [152]
Squamous cell carcinoma	48.3	49.6
Adenocarcinoma	48.4	49.9
SIL	42.4	44.9

Table - 3: Age incidence in cervical carcinoma in study done by Dhakal et al. [152]

AGE (years)	SIL (%)	Squamous-cell carcinoma (%)	Adenocarcinoma (%)
21-30	2.2	---	2.2
31-40	16	48.2	11.1
41-50	35.7	11.1	38.8
51-60	26.7	29.6	35.6
61-70	14.8	11.1	6.7
>70	4.67	---	6.7

Table - 4: Histological subtyping of SCC of cervix

Histological subtyping	Olutoyin G Omoniyi-Esan et al [153]
Large cell non keratinizing SCC	41.1%
Small cell non keratining SCC	9.9%
Micro invasive	1%
Keratinizing SCC	32.3%
Other	15.7%

Table - 5: Incidence of Histological Grading of cervical carcinoma

Grade of tumor	N.Husain et al [154]	E.K.Abudu et al [155]	Swamy et al [156]
Well differentiated	22.8%	39%	35.6%
Moderately differentiated	44.9%	17.7%	47.5%
Poorly differentiated	22.8%	33.3%	16.8%

REFERENCES

1. Sebanti G, Rekah D, Sibani S. A profile of adolescent girls with gynaeco-logical problems. Obstet Gynaecol India 2005; 55(4): 353-5.

2. Mostafa MG, Srivannuboon S, Rachanawutanon M. Accuracy of cytological findings in abnormal cervical smear by cytohistologic comparison. Indian J Pathol Microbiol 2000; 43(10): 23-9.

3. Cannistra SA, Niloff JM.cancer of uterine cervix.N Engi j med 1996; 334: 1030-38.

4. Kristensen GB, Holm R, Abeler VM, Trope CG. Evaluation of the prognostic significance of cathepsin D, Epidermal growth factor receptor, and c-erbB-2 in early cervical squamous cell carcinoma: an immunohistochemical study. Cancer 1996; 78: 433-40.

5. Nair BS, Pillai R.oncogenesis of squamous carcinoma of the uterine cervix. Int Jgynecol pathol 1992; 11: 47-7.

6. Parkin DM, Pisani P, Ferlay J. Global cancer statistics. CA Cancer J Clin 1999; 49(1): 33-64.

7. Berrington de Gonzalez A, Green J. Comparison of risk factors for invasive squamous cell carcinoma and adenocarcinoma of the cervix collaborative reanalysis of individual data on 8,097 women with squamous cell carcinoma and 1,374 women with adenocarcinoma from 12 epidemiological studies. Int J. Cancer 2007; 120(4): 885-91.

8. Ursin G, Peters RK, Henderson BE, d'Ablaing G, Monroe KR, Pike MC. Oral contraceptive use and adenocarcinoma of cervix. Lancet 1994; 344 (8934): 1390-1394.

9. Ricardo DR, Frumovitz M, Milam MR, Edison Capp, Charlotte C. et al.Adenosquamous carcinoma versus adenocarcinoma in early-stage cervical cancer patients undergoing radical hysterectomy: An outcomes analysis Gynecologic oncology 2007;107(3): 458-63.

10. Smith HO, Tiffany MF, Qualls CR, Key CR, The rising incidence of adenocarcinoma relative to squamous cell carcinoma of the uterine cervix in the United States-a 24-year population-based study, Gynecol Oncol 2000;78: 97–05.

11. Grisaru D, Covens A, Chapman B, Shaw P, Colgan T,Murphy J,et al. Does histology influence prognosis in patients with early-stage cervical carcinoma? Cancer 2001; 92(12): 2999-04.

12. Ferlay J, Bray F, Pisani P, Parkin DM.Globocan2000. Cancer incidence, morta- lity and prevanence worlwide.IARCP press: Lyon 2001.

13. World Health Organization (February 2006). "Fact sheet No. 297: Cancer". Retrieved 2010-12-01.

14. Bosch FX, Lorincz A, Munoz N, Meijer CJ, Shah KV.the causal relation between human papillomavirus and cervical cancer.J clin pathol 2002; 55: 244-65.

15. Szarewski A, Cuzick J. Smoking and cervical neoplasia: a review of the evidence.j epidem biostat.1998; 3: 229-56.

16. Parkash V, Carcangiu ML. Uterine papillary serous carcinoma after radiation therapy for carcinoma of cervix. Cancer 1992; 69: 496-01.

17. Fritz A, Percy C, Jack A, Shanmugaratnam K, Sobin LH, Parkin DM, Whelan S. International classification of Diseases for oncology (ICD-0). 3rd edition. World health organization: Geneva 2000.

18. Wright CT, Ferenczy A. Benign diseases of the cervix. In: Kurman RT, editor. Blaustein's pathology of female genital tract. 5th ed. NewDelhi: Springer Verlag 2002; 225-52.

19. Craig P, Lowe D. Non-neoplastic lesions of the cervix. In: Fox H, Well M, editors. Haines and Taylor Obstetrical and Gyanecological pathology 5th ed. Edinburgh Churchill Livingstone 2000; 273-96.

20. Padubidri VG, Daftary SW, editors Howkins and Bourne Shaw's text book of gynaecology. New Delhi: Chruchill Livingstone 2004.

21. Singer A, Chow C. Anatomy of cervix and physiological changes in cervical epithelium. In: Fox H, Wells M, editors. Haines and Taylor obstetrical and gynaecological pathology. Edinburgh: Churchill Livingstone 2003; 5(1): 247-72.

22. Ind T. Female reproductive system. In: Standing S, Ellis H, Healy JC, Johnson D, Williams A, Collins P, Wigley C, editors. Gray's anatomy. The anatomical basis of clinical practice. Edinburgh: Churchill Livingstone 2005; 39: 1331-38.

23. Agar MR, Lee MJ. Pelvis and perineum. In Grant's atlas of anatomy.10th ed. Philadelphia. Lippincott Williams and Wilkins 1999; 169-35.

24. Lawrence DW, Shingleton MH. Early phys iologic squamous metaplasia of the cervix. Light and electron microscopic observations. Am J Surg Pathol 1980; 137: 61.

25. Sankaranarayanan R, Black RB, Parkin DM, Cancer survival in developing countries. Editors. Lyon: IARC Press 1998 (IARC Scientific Publications No. 145).

26. National Cancer Registry Programme. Consolidated report of the Population based Cancer Registries 2001-2004. Indian Council of Medical Research, New-Delhi, 2006.

27. John AR, Howard WJ. Te Linde's operative gynecology Lippincott Williams and Wilkins Publishers 2003; 9:1377.

28. Meisel A, Fortin R. Condylomatous lesion of cervix and vagina, Cytolo-gical patterns.Acta Cytol 1976; 20: 505.

29. Purota E, Savia E. Cytology of Gynecologic Condyloma acuminatum.Acta Cytol1977; 21: 26.

30. ZurHausen. Papillomaviruses in Human, cancer 1987; 59: 1692.

31. Jonathan B. Norvaks Gynecology.published by Lippincott Williams and Wilkins 2003; 13: 478.

32. Rosai and Ackerman's: Surgical pathology 2004; 9(2): 1533.

33. World Health Organization (February 2006). "Fact sheet No. 297: Cancer. http://www.who.int/mediacentre/factsheets/fs297/en/ index. html. Retrieved 2007-12-01.

34. Cancer Research UK website". http://info.cancerresearchuk.org/ cancerstats/ Types/cervix/incidence/.retrieved 2009-1-03.

35. Anderson MC. The cervix, excluding cancer. In: Symmers WST editor. Female reproductive system. Systemic pathology. Edinburgh: Churchill Livingstone 1991; 6: 47-5.

36. Crum CP. Papilloma virus related changes and premalignant and malignant squamous lesions of the uterine cervix. In: Clement PB, Young RH editors. Tumours and tumour like lesions of the uterine corpus and cervix. Roth LM editor. Contemporary issues in surgical pathology 1993; 19: 51-84.

37. Mitta KR, Chan W, Demopoulo SR. Sensitivity and specificity of various morphological features of cervical condylomas. Arch Pathol Lab Med 1990; 114: 1038-42.

38. Fu YS, Reagnan JW. Development anatomy and histology of lower female genital tract. In: Benington JL, editor. Pathology of uterine cervix. Major problems of pathology Philadelphia: WB Saunders 1989; 12 (15).

39. Nucci MR, Fletcher CDM. Fibroepithelial stromal polyp of Vulvovaginal tissue from the banal to the bizarre.pathol case Rev 1998; 3: 151-7.

40. Nucci MR, Young RH, Fletcher CD. Cellular pseudosarcomatous fibroepi-thelial stromal polyp of the lower female genital tract: an underrecognized lesion often misdiagnosed as sarcoma .Am J Surg pathol 2000; 24: 231-40.

41. Prathima KM. Surgical pathology of the uterine cervix in Davangere (unpublished doctoral dissertation), Kuvempu University. Shimoga 1998.

42. Gottlieb SL, Neville AM, Walker F. In: Coleman DV, Evans DMD, editors. Biopsy pathology and cytology of cervix. London: Chapman and Hall 1988; 1:194.

43. Crum CP, Cibas ES, Lee KR. Pathology of early cervical neopasia.New York: Churchill Livingstone 1996.

44. Meisels A, Fortin R. Condylomatous lesion of cervix and vagina.I.cytologic patterns.Acta Cytol 1976; 20: 505-09.

45. Crum CP, Ikenberg H, Richart RM. Human papillomavirus type 16 and early cervical neoplasia.N Engl J Med 1984;310: 880-3.

46. Ward BE, Saleh AM, Williams JV. Papillary immature metaplasia of the cervix: a distinct subset of exophytic cervical condyloma associated with HPV-6/11 nucleic acids.Mod pathol 1992; 5: 391-5.

47. Trivijitsilp P, Mosher RE, Crum CP. Papillary immature metaplasia (immature conyloma) of the cervix: a clinicopatholgical analysis and comparison with papillary carcinoma.Hum Pathol 1998; 29: 641-8.

48. Mosher RE, Lee KR, Trivijitsilp P. Cytologic correlation of papillary immature metaplasia of the cervix.Diagn Cytopathol 1998;18: 416-21.

49. Crum CP. Symposium part I: should the Bethesda system terminology be used in diagnostic surgical pathology (point.)Int J Gynecol pathole 2003; 22: 5-12.

50. Park JJ, Sun D, Quade BJ. Stratified mucin-producing intraepithelial lesion of the cervix: adenosqamous or columnar cell neoplasia Am J Surg Pathol 2000; 24: 1414-19.

51. Padmanabhan H, Oumachigui A, Sankaran V, Rajaram P. A study of 80 cases of cervical intraepithelial neoplasia in a developing country. J Obstet Gynecol India 1998: 107-11.

52. Mittal K, Mesia A, Demopolos RI. MIB-1expression is useful in distinguishing dysplasia from atrophy in elderly women.int JGynecol Pathol 1999; 18: 122-24.

53. Resnick M, Lester S, Tate JE, Sheets EE, Sparks C, Crum CP. Viral and histopathologic correlates of MN and MIB-1 expression in cervical intraepithelial neoplasia.HUM Pathol 1996; 27: 234-39.

54. Keating JT, Cviko A, Riethdorf S, Riethdorf L, Quade BJ, Sun D, Duensing S, Sheets EE, Munger K, Crum CP. ki-67,cyclin E,and p16INK4 are compli-mentary surrogate biomarkers for human neoplasia. AM J surg Pathol 2001; 25: 884-91.

55. Sano T, Oyama T, Kashiwabara K, Fukuda T, Nakajima T Expression status of p16 protein is associated with human papillmavirus oncogenic potential in cervical and genital lesion. AM J Pathol 1998; 153: 1741-1748.

56. Winifred Gray: Diagnostic cytopathology 1st edition. London, person professional Ltd 1995; 3(1): 1352.

57. Coleman DV, Evans D. Biopsy pathology and cytology of cervix. .london Chapaman& Hall 1998; 1: 1251-54.

58. Boyes DA, Worth AJ, Fidler HK.The results of treatment of 4389 cases of preclinical sqamous cell carcinoma. Br J obslet Gynaecol 1973; 77: 769.

59. Savage EW. Microinvasive carcinoma of the cervix. Am J Obslet Gynecol 1972; 113: 708.

60. Nag P, Reagan JW. Microinvasive carcinoma of uterine cervix.Am J Clin Pathol 1969; 52: 511.

61. Robert ME, Fu YS. Squamous cell carcinoma of the uterine cervix a review with emphasis on prognostic factors and unusual variant.Semin Diagn Pathol 1990; 7: 173.

62. Fidler HK, Boyes DA, Worth AJ. Cervical cancer detection in britisg Columbia, Br J Obstet Gynaecol 1968; 75: 392-04.

63. Ylitalo N,Josefsson A,Melbye M,et al.Aprospective study showing long-term infection with human papillomavirus 16 before the development of cervical carcinoma in situ cancer Res 200; 60: 6027-32.

64. Gustafsson I, Adami HO. Naturalhistory of cervical neoplasia: consistent result obtained by an identification technique.Br J cancer 1989; 60: 132-41.

65. Reagan KW, Mamanic MS, Wentz WB. Analytical study of the cells in cervical sqamous cell cancer. Lab Invest 1957; 6: 241-50.

66. Hellweg GD, Doeberitz MV, Marcus J. Trunk. Malignant epithelial tumor. Gabriele Schröder, Heidelberg .Color Atlas of Histopathology of the Cervix Uteri, Germany, Springer-Verlag Berlin Heidel-berg 2006; 2: 134-35.

67. Pratt DW, Lee SC. Verrucous carcinoma of the cervix.Am J Obstet Gynecol 1977; 129: 699.

68. Toki T, Kurman RJ, Park JS, Kessis T, Daniel RW, Shah KV probable non-papillomavirus etiology of sqamous cell carcinoma of the vulva in older women;a clinicopathologic study using in situ hybridization and polymerase chain reaction.Int J Gynecol Pathol 1991;10: 107-25.

69. Lininger RA, Wistuba I, Gazdar A, Koenig C, Tavassoli FA, Albores-Saavedra J Human papilloma virus type 16 is dectected in transitional cell carcinoma and squamotrasitional cell carcinoma of cervix and ecdometrium. Cancer 1998; 83: 521-27.

70. Maitra A, Wistuba, Gibbons D, Gazdar AF, AlboresSaavedra J. Allelic losses at chromosome 3p are seen in human papillom virus 16 associated transit-ional cell carcinoma of the cervix. Gynecol Oncol 1999; 74: 361-68.

71. Steeper TA, Piscioli F, Rosai J.Sqamous cell carcinoma with sarcoma like stroma of the female genital tract: clinicopathological study of four cases .cancer 1983; 52: 890-96.

72. Cotran RS, Kumar V, Robins SL. ln: Female Genital Tract. Robins Pathological basis of disease, Philadelphia. W. B. Saunders Company 1989; 4: 1127-80.

73. Female reproductive system uterus cervix. In: Rosai J (ed.) Ackennans surgical pathology. Washington D.C. C. v. Mosby Company 1989; 7: 997-1191.

74. Aridogen N, Cetin T, Kadayifci O. Giant cervical polyp due to a foreign body in a virgin. Aust N Z J Obstet Gynaecol 1988; 28: 146–7.

75. Wright CT, Ferenczy A. Benign diseases of the cervix. In Kurman RT, editor. Blaustein's pathology of female genital tract. NewDelhi: Springer Verlag 2002; 5: 225-52.

76. Craig P, Lowe D. Non-neoplastic lesions of the cervix. In: Fox H, Well M, editors. Haines and Taylor Obstetrical and Gyanecological pathology Edin-burgh: Churchill Livingstone 2003; 5: 273-96.

77. Gottlieb SL, Neville AM, Walker F. In: Coleman DV, Evans DMD, editors. Biopsy pathology and cytology of cervix. London: Chapman and Hall 1988; 1-194.

78. Crum CP, Nucci MR, Lee KR. The cervix. In: Mills SE, Carter D, Greenson JK, Oberman HA, Reuter V, Stoler MH, editors. Sternberg's diagnostic surgical pathology. Philadelphia: Lippincott Williams and Wilkins 2004; 47(3): 2377-434.

79. Andrews CF, Jourdain L, Damjanov I. Benign cervical mesonephric papiloma of childhood: report of a case studied by light and electron microscopy.Diagn Gynecol Obstet 1981; 3: 39-43.

80. Christopherson WM, Nealon N, Gray LA. Noninvasive precursor lesion of adenocarcinoma and mixed adenosqamous carcinoma of the cervix uteri .Cancer 1979; 44: 975-83.

81. Andersen ES, Arffmann E. Adenocarcinoma in situ of the uterine cervix a clinicopathological study of 36 cases.Gynecol Oncol 1989; 35: 1-7.

82. Ayer B, Pacey F, Greenberg M. The cytologic diagnosis of adenocarcinoma in situ of the cervix uteri and related lesion. I. Adenocarcinoma in situ.Acta Cytol 1987; 31: 391-11.

83. Betsill WL, Clark AH. Early endocervical glandular neoplasia.I. Histomor-phology and cytomorphology. Acta Cytol 1986; 30: 115-25.

84. Lee KR, Manna EA, Jones MA.Comparative cytology feature of adenocarci-noma in situ of the uterine cervix. Acta Cytol 1991; 35: 117-26.

85. Quizilbash AH. In situ and microinvasive adenocarcinoma of the uterine cervix: a clinical cytology and histology study of 14 cases. Am J Clin Pathol 1975; 64: 155-70.

86. Piirog EC, Kleter B, Olgac S. Prevalence of human papillmavirus DNA in different histological subtype of cervical adenocarcinoma.Am J Pathol 2000; 157: 1055-62.

87. Biscotti CV, Hart WR. Apoptotic bodies: a consistent morphological feature of endocervical adenocarcinoma in situ.Am J Surg Pathol 1998; 22: 434-39.

88. Webb JC, Key CR, Qualls CR. Papulation-based study of microinvasive adenocarcinoma of uterine cervix.Obstet Gynecol 2001; 701-06.

89. Schorge JO, Lee KR, Sheets EE. Prospective management of stage IA Cervical adenocarcinoma by conisation alone to preserve fertility.Gynecol Oncol 2000; 78: 217-20.

90. Nicklin JL, Perrin LC, Crandon AJ. Microinvasive adenocarcinoma of the cervix Aust NZ J Obstet Gynaecol 1999; 39: 411-13.

91. Ostor AG. Early invasive adenocarcinoma of uterine cervix. Int J Gynecol Pathol 2000; 19: 29-38.

92. Miller BE, Flax SD, Arheart K. The presentation of adenocarcinoma of the uterine cervix. Cancer 1993; 72: 1281-85.

93. Vesterinen E, Forss M, Nieminen U.Increase of cervical adenocarcinoma: A report of 520 cases of cervical carcinoma including 112 tumor with glandular element.Gynecol Oncol 1989; 33: 49-53.

94. Ursin G, Pike MC, Preston-martin S. sexual, reproductive, and other risk factors for adenocarcinoma of the cervix: result from a population–based case control study.Cancer causes control 1996; 7: 391.

95. Thomas DB, Ray RM. Oral contraceptive and invasive adenocarcinoma and adenosqamous carcinoma of uterine cervix (the World Health Organization collaboration collaborative study of neoplasia and steroid contraceptives). Am J Epidemol 1996; 144: 281-89.

96. Brinton LA, Tashima KT, Lehman HG. Epidemiology of cervical cancer by cell type.Cancer Res 1987; 47: 1706-11.

97. Young RH, Clement PB, Scully RE.Premalignant and malignant glandular lesion of the uterine cervix.In Clement PB,Young RH,eds.Tumor and tumor-like lesion of the uterine corpus and cervix. New York: Churchill Livingstone, 1993.

98. Eifel PJ, Burke TW, Morris M. Adenocarcinoma as an independent risk factor for disease recurrence in patients with stage IB cervical carcinoma. Gynecol Oncol 1995; 59: 38-44.

99. Vega de la G. Signet ring cell carcinoma of the uterine cervix, Patologia 1976; 14: 193-96.

100. Gilks CB, Young RH, Aguirre P. Adenoma malignum (minimal deviation adenocarcinoma) of the uterine cervix:a clinicopathologic and immunohistochemical analysis of 26 cases.Am J Surg Pathol 1989; 13: 717-29.

101. McGowan L, Young RH, Scully RE. Peutz-jeghers syndrome with "adenoma malignum"of the cervix: a report of two cases. Gynecol Oncol 1980; 10: 125-33.

102. Rahilly MA, Williams ARW, Al-Nafussi A.Minimal deviation endometrioid adenocarcinoma of the cervix: a clinicopathologic and immunohistochemical study of two cases. Histopathology 1992; 20: 351-54.

103. Scully RE, Welch WR. Pathology of the female genital tract after prenatal exposure to diethylstilbestrol.In Herbst AI, Bern HA, eds.Development effect of Diethylstibestrol (DES) in pregnancy. New York: Thieme-stratton 1981; 26.

104. Nordqvist SRB, Fidler W Jr, Woodryff J. Clear cell adenocarcinoma of the cervix and vagina: a clinicopathologic study of 21 cases with and without a history of maternal ingestion of estrogens.cancer 1976; 37: 858-71.

105.   Kaminski PF, Maier RC. Clear cell adenocarcinoma of the cervix unrelated to diethylstilbestrol exposure.Obstet Gynecol 1983; 62: 720-27.

106.   Clement PB, Young RG, Keh P. Malignant mesonephric neoplasm of the uterine cervix:a report of eight cases,including four with a malignant spindle-cell component.Am J Surg Pathol 1995; 19: 1158-71.

107.   Ordi J, Nogales FF, Palacin A. Mesonephric adenocarcinoma of the uterine corpus: CD10 expression of mesonephric differentiation.Am J Surg pathol 2001; 25: 1540-45.

108.   Silver SA, Devouassoux-shisheboran M, Mezzetti TP. Mesonephric adenocarcinoma of the uterine cervix: a study of 11 cases with immune-histochemical finding.Am J Surg Pathol 2001; 25: 379-87.

109.   Benda J, Platz C, Buchsbaum H, Lifshitz S. Mucin production in defining mixed carcinoma of the uterine cervix: a clinicopathologic study. Int J Gynecol Pathol 1985; 4: 314-27.

110.   Hopkins MP, Schmidt RW, Roberts JA. The prognosis and treatment of stage I adenocarcinoma of the cervix. Obstet Gynecol 1988; 72: 915-21.

111.   Look KY, Brunetto VL, Clarke-pearson DL. An nalysis of cell type in patients with surgical oncol 1996; 63: 304-11.

112.   Littman P, Clement PB, Henriksen B. Glassy cell carcinoma of the cervix. Cancer 1976; 37: 2238-46.

113.   Ferry JA, Scully RE. Adenoid cystic carcinoma and adenoid basal carcinoma of the uterine cervix: a study of 28 cases. Am J, Surg pathol 1988; 12: 134-44.

114.   Albores-Saavedra J, Gersell D, Gilks CB, Henson DE, Lindberg G, Santiago H, Scully RE, et al. Terminology of endocrine tumor of the uterine cervix: result of a workshop sponsored by the college of American pathologists and the national cancer institute.Arch pathole Lab Med 1997; 121: 34-39.

115.   Abeler VM, Holm R, Nesland JM, Kjorstad KE. Small cell carcinoma of the cervix.A clinicopathologic study of 26 patient's Cancer 1994; 73: 672-77.

116. Abell MR, Ramirez JA. Sarcomas and carcinosarcoma of the uterine cervix. Cancer 1973; 31: 1176-92.

117. Keel SB, Clement PB, Prat J, Young RH. Malignant schwannoma of the uterine cervix; a study of three cases, Int J Gynecol Pathol 1998; 17: 223-30.

118. Daya DA, Scully RE. Sarcoma botryoides of the uterine cervix in young women: a clinicopathological study of 13 cases, Gynecol Oncol 1968; 29: 290-04.

119. Balat O, Balat A, Verschraegen C, Tomos C, Edwards CL. Sarcoma botryoides of the uterine endocervix: long term result of conservative surgery, Eur J Gynaecol Oncol 1996; 17: 335-37.

120. Gray GF, Glick AD, Kurtin PJ, Jone HW. Alveolar soft part sarcoma of the uterus. Hum Pathol 1986; 17: 297-300.

121. Nielsen GP, Oliva E, Young RH, Rosenberg AE, Dickersin GR, Scully RE. Alveolar soft part sarcoma of the female genital tract:a report of nine cases and review of the literature . Int J Gynecol patho 1995; 114: 283-92.

122. Schammel DP, Tavassoli FA. Uterine angiosarcomas: a morphologic and immunehistochemical study of four cases. Am J Surg Pathol 1998; 22: 246-50.

123. Junge J, Horn T, Bock J. Primary malignant schwannoma of the uterine cervix .case report. Br J Obstet Gynaecol 1989; 96:111-16.

124. Keel SB, Clement PB, Prat J, Young RH. Malignant schwannoma of the uterine cervix:a study of three cases. Int J Gynecol Pathol 1998; 17: 223-30.

125. Nielsen GP, Young RH. Mesenchymal tumors and tumor like lesion of the femailgenital tract; a selective review with emphasis on recently descrived entities. Int J Gynecol Pathol 2001; 20: 105-27.

126. Tiltman AJ. Leiomyomas of the uterine cervix; a study of frequency, Int J Gynecol Pathol 1998; 17: 231-34.

127. Disant Agnese PA, Knowles DM. Extracardiac rhabdomyoma; a clinico-pathological study and review of the literature. Cancer 1980; 46: 780-89.

128. Nielsen GP, Young RH Mesenchymal tumors and tumor like lesion of the femailgenital tract; a selective review with emphasis on recently descrived entities. Int J Gynecol Pathol 2001; 20: 105-27.

129. Kay S, Schneider V. Reactive spindle cell nodule of the endocervix simul-ating uterine sarcoma. Int J Gynecol Pathol 1985; 4: 255-57.

130. Clement PB, Zubovits JT, Young RH, Scully RE. Malignant mullerian mixed tumors of the uterine cervix;a report of nine cases of a neoplasm with morph-ology often different from its counterpart in the corpus. Int J Gynecol Pathol 1998; 17: 211-22.

131. Grayson W, Taylor LF, Cooper K Carcinosarcoma of the uterine cervix; a report of eight cases with immunohistochemical analysis and evaluation of human papillomavirus status. Am J Surg Pathol 2001; 25: 338-47.

132. Babin EA, Davis JR, Hatch KD, Hallum AV. Wilms tumor of the cervix; a case report and review of the literature. Gynecol Oncol 2000; 76: 107-11.

133. Benatar B, Wright C, Freinkel AL, Cooper K. Primary extrarenal wilms tumor of the uterus presentation as a cervical polyp. Int J Gynecol Pathol 1998; 17: 277-80.

134. Young RH, Scully RE. Ovarian sertoli cell tumors; a report of 10 cases. Int J Gynecol Pathol 1984; 2: 349-63.

135. Cantuaria G, Angioli R, Nahmias J, Estape R, Penalvver M. Primary malignant melanoma Of the uterine cervix;case report and review of the literature. Gynecol Oncol 1999; 75: 170-74.

136. Furuya M, Shimizu M, Nishihara H, Ito T, Sakuragi N, Ishikura H, Yoshiki T. Clear cell variant of malignant melanoma of the uterine cervix a case report and review of the literature. Gynecol Oncol 2001; 80: 409-12.

137. Jones HW, Droegemueller W, Makowski EA. A primary melanocarcinoma of the cervix. Am J Obstet Gynecol 1971; 111: 959-63.

138. Uehara T, Izumo T, Kishi K, Takayama S, Kasuga T. Stromal melanocytic foci['blue nevus'] in step section of the uterine cervix. Acta Pathol Jpn 1991; 41; 751-56.

139. Uehara T, Takayama S, Takemura T, Kasuga T. Foci of stromal melanocytes (so-called blue naevus) of the uterine cervix in Japanese women.Virchows Arch. Pathol Anat Histopathol 1991; 418: 327-31.

140. Marsh DJ, Kum JB, Lunetta KL, Bennett MJ, Gorlin RJ, Ahmed SF, Bodurtha J. PTEN mutation spectrum and genotype-phenotype corre-lation in Bannayan Riley-Ruvalcaba syndrome sugest a single entity with cowden syndrome. Hum Mol Genet 1999; 8: 1461-72.

141. Copeland LJ, Sneige N, Ordonez NG, Hancock KC, Gershenson DM, Saul PB, Kavanagh JJ. Endodermal sinus tumor of the vagina and cervix. Cancer 1985; 55: 2558-65.

142. Mauz-Korholz C, Harms D, Calaminus G, Gobel U. Primary chemotherapy and conservative surgery for vaginal yolksac tumor. Maligne Keimzell -tumoren study group. Lancet 2000; 355: 625.

143. Khoor A, Fleming MV, Purcell CA, Seidman JD, Ashton AH, Weaver DL .Mature teratoma of the uterine cervix with pulmonary differentiation. Arch Pathol Lab Med 1995; 119: 848-50.

144. Harris NL, Scully RE. Malignant lymphoma and granulocytic sarcoma of uterus and vagina. A clinicopathologic analysis of 27 cases. Cancer 1984; 53: 2530-254.

145. Sebanti G, Rekah D, Sibani S. A profile of adolescent girls with gynaeco-logical problems. Obstet Gynaecol India 2005; 55(4):353-5.

146. Mostafa MG, Srivannuboon S, Rachanawutanon M. Accuracy of cytological findings in abnormal cervical smear by cytohistologic comparison. Indian J Pathol Microbiol 2000; 43(10):23-9.

147. Cannistra SA, Niloff JM.cancer of uterine cervix.N Engi j med 1996; 334:1030-38.

148. Kristensen GB, Holm R, Abeler VM, Trope CG. Evaluation of the prognostic significance of cathepsin D, Epidermal growth factor receptor, and c-erbB-2 in early cervical squamous cell carcinoma: an immunohistochemical study. Cancer 1996; 78:433-40.

149. Nair BS, Pillai R.oncogenesis of squamous carcinoma of the uterine cervix. Int Jgynecol pathol 1992; 11:47-7.

150. Parkin DM, Pisani P, Ferlay J. Global cancer statistics. CA Cancer J Clin 1999; 49(1):33-64.

151. Berrington de Gonzalez A, Green J. Comparison of risk factors for invasive squamous cell carcinoma and adenocarcinoma of the cervix collaborative reanalysis of individual data on 8,097 women with squamous cell carcinoma and 1,374 women with adenocarcinoma from 12 epidemiological studies. Int J. Cancer 2007; 120(4):885-91.

152. Ursin G, Peters RK, Henderson BE, d'Ablaing G, Monroe KR, Pike MC. Oral contraceptive use and adenocarcinoma of cervix. Lancet 1994; 344 (8934):1390-1394.

153. Ricardo DR, Frumovitz M, Milam MR, Edison Capp, Charlotte C. et al.Adenosquamous carcinoma versus adenocarcinoma in early-stage cervical cancer patients undergoing radical hysterectomy: An outcomes analysis Gynecologic oncology 2007;107(3):458-63.

154. Smith HO, Tiffany MF, Qualls CR, Key CR, The rising incidence of adenocarcinoma relative to squamous cell carcinoma of the uterine cervix in the United States-a 24-year population-based study, Gynecol Oncol 2000;78: 97–05.

155. Grisaru D, Covens A, Chapman B, Shaw P, Colgan T,Murphy J,et al. Does histology influence prognosis in patients with early-stage cervical carcinoma? Cancer 2001; 92(12):2999-04.

156. Ferlay J, Bray F, Pisani P, Parkin DM.Globocan2000. Cancer incidence, morta- lity and prevanence worlwide.IARCP press: Lyon 2001.